EMPEROR T

CW00498077

How to Care for Your Monkey

Monkey Training Book for Beginners From Background, Appearance, Behaviors, Feeding, Health and breeding + Much More

NEMY LULEE

Contents

CHAPTER ONE

Emperor Tamarin

One of the most interesting and well-known primates in the world is the Emperor Tamarin. The Amazon Rainforest in South America is home to Emperor Tamarin. Its long, white mustache, which lends it its name, makes it simple to identify.

Emperor Tamarins are tiny primates that may weigh up to 1.5 pounds and have a body length of up to 12 inches. They have a white mustache that often protrudes from the sides of their face and reddish-brown fur on their body.

They also have long tails that are useful for balance, as well as black and white stripes on their faces and chests. The Emperor Tamarins spend most of their time in trees since they are arboreal animals. They coexist in tiny family units of two to six members, each of which includes a dominant male, multiple females, and their offspring.

They are omnivorous, meaning they consume both plants and animals, although they choose fruit, insects, and small vertebrates as food sources.

The daytime is when Emperor Tamarins are most active, and they are known to communicate by using loud vocalizations. They have a distinctive vocalization that they use to draw attention to themselves and utilize their lengthy mustaches to detect their surroundings.

The population of Emperor Tamarin is now steady, and it is a crucial species for its natural environment. Deforestation, which alters their habitat, and pet trade hunting are threats to them. To safeguard their habitat and guarantee their continued existence in the wild, conservation initiatives are being done.

A well-known species, Emperor Tamarin has become a favorite in literature, media, and entertainment. Because of its unusual look and behavior, it is an enthralling topic to study, and conservation efforts are ensuring that future generations may appreciate this wonderful species.

Background and Habitat

The tropical rainforest of South America is home to the Emperor Tamarin (Saguinus imperator), a kind of tamarin monkey. It is indigenous to Bolivia, Brazil, Peru, and the western Amazon Basin. The species' long white mustache, which is supposed to be the origin of its name, is well recognized for it.

The Emperor Tamarin is a medium-sized monkey that may weigh up to 2.2 lbs. and has a body length of 11.5 to 14.5 inches. They resemble a mustache because of their long snout and characteristic white hair tuft on the top lip.

They have a black body with white markings on the chest, neck, and sides of the face. The Emperor Tamarin has thick, dense fur that is normally gray with a rusty undertone.

Being a diurnal animal, the Emperor Tamarin is awake during the day and asleep at night. It makes its home in the rainforest's canopy, where it hunts for food and makes nests. It is a sociable animal that often gathers in groups of two or more people.

Being an arboreal species, the emperor tamarin spends most of its time in trees. In pursuit of food, it climbs and leaps from branch to branch. Its major food sources are fruits, flowers, and insects. It will sometimes consume small vertebrates like frogs, lizards, and snakes.

The humid lowland woods of the western Amazon Basin are home to Emperor Tamarin. Brazil, Peru, and Bolivia are the nations where it is present. It favors secondary-growth woods that have seen human disturbance, including logging.

Additionally, it may be found in palm forests and gallery forests, which are stretches of trees along rivers and streams. The Emperor Tamarin is a vocal species that uses a range of sounds to communicate with its group. It converses with other members of its species by growing a lengthy mustache.

When a tamarin is in danger or getting ready to mate, for example, the mustache is utilized to alert other tamarins. Due to habitat degradation and poaching, the Emperor Tamarin is a vulnerable species.

There are said to be between 5000 and 10000 people living there. To save this species and its environment, conservation measures are being made. The Tambopata-Candamo National Reserve in Peru and the Yasun National Park in Ecuador are two places where the species is protected.

Appearance and Behaviors

A little species of monkey found only in Bolivia, Brazil, and Peru's southwest Amazon Basin is known as the Emperor Tamarin. It has recognizable facial characteristics, and both its appearance and behavior are distinctive.

Appearance The Emperor Tamarin is a little monkey with a tail that may grow up to sixteen inches long. Its usual length is between eight and twelve inches.

Its hair is typically grey in hue with a black muzzle, and white fur tufts on each side frame the face. The Emperor Tamarin gets its name from its long, white mustache, which is its most distinctive characteristic.

The Emperor Tamarin's long, bushy tail serves as a means of balance as it leaps from branch to branch. The Emperor Tamarin is a gregarious creature that lives in packs of up to twenty people. The gang would often hang around together all day while playing and scavenging for food. The alpha male establishes the direction of the group, and the other males will follow.

In addition, the alpha male will defend the territory of the group and safeguard it from predators. Emperor Tamarin mostly consumes fruits and insects, and it often uses its long tail to get food in awkward locations. It can find food because of its acute hearing and smelling senses.

Emperor Tamarin is a talkative animal, and it communicates with its group by using calls. It is a quick climber as well, and the long tail aids in maintaining balance when it jumps from branch to branch.

14

The lively Emperor Tamarin will spend most of the day investigate its environment. It also loves engaging with the other members of its group and is fun and gregarious. Conclusion With a distinctive facial structure and a variety of behaviors, the Emperor Tamarin is a rare and intriguing species of monkey.

It is an easily recognized species because of its long white mustache and bushy tail, and it has remarkable social and vocal behavior. The Emperor Tamarin may be an interesting and enjoyable pet with the right care and surroundings.

Communication and Handling

The Emperor Tamarin is a little species of monkey that is indigenous to the rainforests of South America. It gets its name from the long, silky white hair that resembles a mustache. When caring for Emperor Tamarins, it is vital to think about handling and communication.

First of all, Emperor Tamarin relies heavily on communication. They depend on speech to engage and build connections with other people since they are social animals.

They communicate via a range of vocalizations, gestures, and facial and body language. Short, high-pitched chirping sounds are made during vocalizations to notify other animals of the presence of food, danger, or possible mates.

From violence and terror to submission and humor, body language and facial expressions may convey a wide variety of feelings and intentions.

Understanding and interpreting Emperor Tamarins' diverse communication methods is crucial for efficient communication. This may be achieved by paying attention to the various behaviors kids display in various situations and taking appropriate action.

For instance, if an Emperor Tamarin approaches you with its mouth open and ears back, this might be an aggressive sign, and you should proceed cautiously. On the other hand, if an Emperor Tamarin approaches you while keeping its mouth shut and ears perked, this might be a sign of interest and should be handled gently.

The handling of Emperor Tamarins is a crucial component of their care. This species is inherently apprehensive of people and might grow agitated in the wrong circumstances.

Because of this, it is critical to treat them with care and gentleness. Always use both hands while holding an Emperor Tamarin, one to support the body and the other to support the head and neck. Never move quickly or aggressively; instead, maintain all motions gradually and deliberately.

It is also crucial to speak to Emperor Tamarin in a soothing, quiet tone and to refrain from making abrupt movements or loud sounds. Finally, it is important to be mindful of

Emperor Tamarins' innate habits and mannerisms while managing them.

Being inherently timid, this species may flee or hide if it perceives danger. The best course of action in this situation is to leave the animal alone and wait for it to emerge on its own. In conclusion, handling and communication are crucial issues to think about while taking care of Emperor Tamarins.

It is possible to found effective communication with these intelligent and sociable animals by learning to identify in addition to identify in the midst of their vocalizations, facial expressions, in addition to body language, as well as by treating them in the midst of care in addition to gentleness.

CHAPTER TWO

Socializing and Aggression

The South American Amazon rainforest is home to the little monkey species known as the Emperor Tamarin (Saguinus Imperator). It is one of the bigger kinds of tamarin and is distinguished by its spectacular mustache.

The gregarious and energetic Emperor Tamarin lives in groups of up to eight other people. They frolic and groom each other while they seek food together. A significant aspect of Emperor

Tamarin's existence is social interaction.

They develop strong ties with their tribe members since they are sociable creatures by nature. The tamarins often roam in packs, remaining close together as they interact, play, and groom one another. Additionally, they exhibit bonding behavior by sharing a bed.

This social activity is believed to assist tamarins to defend themselves against predators and preserve their dominant hierarchies. The Emperor Tamarin has a complex mating strategy that includes social interaction.

They are polyandrous, which means that each female in a flock will mate with different males. As the males vie for the attention of the mother, this enhances the likelihood of successful offspring. The males' ability to build dominance hierarchies via social interaction will decide which guys are most effective at mating.

Aggression, however, may also result from social interaction. The Emperor Tamarin is well renowned for its combative tendencies, especially while protecting its territory.

Tamarins will chase, bite, and scratch intruders away from their food supplies or nesting sites using vocalizations and physical assaults like these. When vying for mates, they will also behave aggressively.

The interactions of Emperor Tamarin with other species also exhibit aggression. They have been known to attack marmosets and other monkeys, as well as birds, lizards, and snakes. Aggression is seen as a means of establishing dominant hierarchies and protecting one's territory and food supplies. Emperor Tamarin is a gregarious and energetic creature that relies heavily on both aggressiveness and social interaction to survive.

Aggression is needed to defend one's territory and food supplies while socializing is crucial for sustaining dominance hierarchies. knowledge of Emperor Tamarin and its role in the Amazon jungle requires knowledge of their behavior.

Breeding Procedure In Steps

Finding a suitable Pair is the First Step in producing Emperor Tamarins

1: Finding a Suitable Pair

Typically, this entails pairing up two members of the same species with opposing sexes. The chance of successful mating is increased while searching for a couple by choosing animals with comparable traits and habits.

In order to guarantee that the progeny is healthy, it is also crucial to seek animals that are in excellent health. In order to avoid any territorial or aggressive behavior, it is preferable to introduce the couple in a neutral setting.

2: Establishing a Cage

Following the selection of an Emperor Tamarin couple, the following step is to create an ideal nesting habitat. This should be a safe, cozy space that is designed to provide shelter and warmth, like a cage or an aviary. The animals will be able to unwind and concentrate on reproducing if the location is free from interruptions and distractions.

3: Providing Healthy Food

Emperor tamarins need a portion of healthy food in order to

successfully reproduce. A vitamin and mineral supplement should also be included, as well as a range of fruits, vegetables, and insects. The animals' health will be preserved if the food is balanced and comprehensive, thus it is crucial to make sure of this.

4: Monitoring the Pair

After the nesting habitat has been created and the pair has been formed, it is crucial to keep a constant eye on the animals to ensure that mating is successful. This entails watching the animals' behavior and activities and looking out for any indications of hostility or discomfort. In order to prevent any possible troubles, it is crucial to treat any issues as soon as they appear.

5: Caring for the Offspring

After successful mating and the birth of the offspring, it is crucial to provide them with the finest care possible. This entails offering a hygienic and cozy atmosphere as well as balanced food. Additionally, it is crucial to keep a constant eye on the young animals to make sure they are healthy and developing normally.

Emperor tamarin breeding may be a pleasant and fulfilling endeavor. By doing these actions, you may improve your chances of success and make sure the children are robust and healthy. Your Emperor Tamarins may grow to their maximum potential with the right care and attention.

Feeding

The Amazon jungle is home to the very rare species of monkey known as the Emperor Tamarin. As a consequence, it is critical to provide a portion of food that satisfies its nutritional requirements and is suitable for its surroundings.

If the right procedures are followed, feeding an Emperor Tamarin is a surprisingly easy activity that may be completed with little effort. The Emperor Tamarin's main food sources include fruits, flowers, leaves, and insects.

Fruits and flowers should make up the bulk of their diet since they are this species' primary sources of sustenance. If you want to feed an Emperor Tamarin, oranges, apples, mangoes, and bananas are all great options.

Since Tamarin's mouth is too tiny to accommodate bigger portions, these fruits should be sliced into little pieces.

The Emperor Tamarin should be fed a variety of foliage, flowers, and insects in addition to fruits. Small, freshly cut pieces of leaves from a range of various trees should be supplied.

Flowers should be presented as well, but they should be ones that are indigenous to the rainforest, including bromeliads, orchids, and heliconias. Crickets are a fantastic option for insects since they are easily accessible and a source of protein.

It is critical to feed an Emperor Tamarin food that is both fresh and devoid of poisons and pesticides. Before feeding, fruits should be carefully cleaned and checked for indications of rotting. Check the leaves and blossoms for any indications of contamination or damage as well.

The Tamarin will obtain a balanced diet if you supply a range of meals, which is another crucial consideration. The Tamarin must also have access to fresh water since they are very sensitive to dehydration. Water should be provided in a small basin or dish and should be replaced every day.

In order to enhance Tamarin's water intake, it is advantageous to provide fruits high in water content like oranges and cantaloupe. Finally, it is important to keep an eye on Tamarin's diet and well-being.

Any indications of disease or malnutrition should be kept an eye out for in the tamarind, and if any are found, a veterinarian should be contacted right once. Additionally, it could be important to provide the tamarin with various kinds of food if it is not eating in order to get it to start.

If the right instructions are followed, feeding an Emperor Tamarin is a rather easy process. The Tamarin will eat a complete diet if fresh fruits, leaves, flowers, and insects are provided.

Fresh water should also constantly be available, and Tamarin should be watched closely for any indications of illness. Emperor Tamarin can grow and live a long, happy life in its native environment with the right care and nourishment.

Health care and Disease

Two of the most significant topics in the study of Emperor Tamarins are healthcare and sickness. Small, omnivorous primates called Emperor Tamarins are located in South America's Amazon jungle.

They are distinguished by their striking black and white fur as well as their long, white facial hair that resembles a mustache. A wide range of fruits, flowers, leaves, insects, and tiny vertebrates make up the Emperor Tamarin's diet. Additionally, they eat a variety of water-soaked fruits, flowers, and leaves.

This enables them to get the nutrients and moisture they need to live. Emperor Tamarins are also known to consume other animals' excrement, which might be contaminated with various germs and parasites.

The Emperor Tamarin is unfortunately susceptible to a wide range of illnesses and health issues. They are susceptible to common ailments such as skin infections, diarrhea, and upper respiratory infections. Additionally, parasites and other infectious disorders might affect them.

Additionally, because of their diet, they may have dehydration and nutritional deficits. Emperor Tamarins must get the right medical attention in regulate to sustain their healthiness as well as well-being. This entails giving them well-balanced food, a tidy and welcoming home, and frequent veterinarian checkups.

Additionally, it is critical to keep a careful eye on their health since any changes in their behavior or looks might be a sign of a larger problem. It is essential to take care of any possible illnesses and infections in addition to providing routine medical treatment.

Any potential parasites or diseases should be found and treated as soon as feasible. Antibiotics may sometimes be required to treat certain illnesses. The Emperor Tamarins should have regular vaccines to safeguard them against common infections and illnesses.

The public should be made aware of the requirements of the Emperor Tamarins and encouraged to support the preservation of their habitats. We can guarantee that these rare primates enjoy long and healthy lives by raising awareness about the value of protecting the rainforest and its species. In conclusion, the two most significant topics in the research of Emperor Tamarins are health care and illness.

These primates can live long, healthy lives with the right medical treatment and illness prevention. We can support the Emperor Tamarins in the Amazon jungle in being healthy and thriving by giving them a balanced meal, a tidy and pleasant living space, and regular veterinarian appointments.

CHAPTER THREE

Exercise and Training

The Emperor Tamarin is a very active and vivacious type of monkey, thus it is crucial to provide them with enough exercise and training to be strong and content.

Emperor Tamarins need to exercise and train to be healthy, thus it is important to practice moderation and make informed decisions while doing so. Emperor Tamarin should exercise every day, and it should be done in a method that is both secure and age-appropriate.

Tamarins are quick and nimble animals who can run, leap, and climb with ease. They have to be permitted to do so in a secure enclosure, like a sizable playpen or cage.

Their inherent interest and agility will be stimulated by a range of activities, which will keep them busy and healthy. Running up and down ramps, leaping from platforms, and swinging from ropes are some of these activities.

Emperor Tamarins should have plenty of opportunities for cerebral stimulation in addition to physical exercise.

Swings, rope ladders, and puzzle toys are excellent methods to keep their brains active and provide them with entertainment when they become bored.

Emperor Tamarins benefit from training since it teaches them how to communicate with people and other animals. Tamarins are naturally wary and may quickly get overwhelmed, so training should be done gradually and patiently.

Treats should be given as rewards for excellent conduct since positive reinforcement is essential. The Tamarin should be trained to understand its name and the fundamental commands like sit, stay, and come.

An Emperor Tamarin's daily regimen should include exercise and training, which should be done carefully and in moderation. Keeping them active and mentally engaged will keep them happy and healthy, and teaching them how to communicate with people and other animals will make them feels more at home in their surroundings.

Emperor Tamarins may live long, healthy lives with the correct amount of training and activity.

Housing and Grooming

Small primates called emperor tamarins are groomed and housed in their native South America. With a black face, orange fur, and a white mustache, they have a unique and stunning physical look that has become well-known.

Since tamarins are gregarious creatures, keeping them in captivity successfully requires specialized care. In order to do this, sufficient housing and grooming must be provided.

When it comes to emperor tamarin housing, it is important to provide them with a cage that is roomy enough for them to roam about freely and have plenty of room to explore and play. Additionally, the cage needs to be able to provide them with a safe and cozy place to live.

A reasonable rule of thumb is to provide each tamarin with at least four square feet of room. For instance, you should provide your two emperor tamarins with a cage that is at least eight square feet in size. To encourage exploration and easy movement, the cage should not only be large but also contain several levels, such as platforms or shelves.

A peaceful part of the house, away from any activity or loud noise, should also be chosen for the cage's location. The cage should also be equipped with perches so that the tamarins have somewhere to play, climb, and leap.

In addition to giving your emperor tamarins proper shelter, it is critical to provide them with routine grooming. This includes cutting their nails and combing their hair to get rid of any mats and knots. This must be done often to maintain the tamarins' healthy nails and maintain the quality of their hair.

Giving your tamarins a bath once every two weeks is also crucial. This aids in keeping their fur clean and dirt- and parasite-free. It is crucial to bathe your tamarins in lukewarm water with a mild, unscented soap.

To ensure that all of the soap is gone, it is also crucial to properly rinse them. Proper housing and grooming are crucial for maintaining the health and happiness of your emperor tamarins. For them to go around and explore, a large cage with several levels and platforms must be provided.

To keep their hair clean and clear of parasites or filth, frequent brushing, and washing are also essential. Your emperor tamarins will be guaranteed to flourish in captivity with the proper care.

Male Versus Female

The South American Emperor Tamarin is a tiny species of monkey that is well-known for looking somewhat like a human emperor. Male and female members of the species are split into two separate sexes. They are intriguing animals to examine since both genders differ from one another in certain traits and behavior.

The average male Emperor Tamarin weighs between 500 and 600 grams, and the average female weighs between 400 and 500 grams. Additionally, the males have a stunning silver mane that extends around the back of their heads, longer hair, and more vibrant markings.

They are also more aggressive when interacting with other tamarins and have bigger canines than their female counterparts. Emperor Tamarin females are smaller and more delicate-looking than males.

She has less colorful, shorter hair, and a less noticeable mane. Additionally, females often exhibit less aggression and have smaller canines. More time is spent looking for food and socializing with other monkeys by male Emperor Tamarins than by females.

When protecting their area, they are more aggressive and often confront other males for supremacy. Females, on the other hand, are far less active and prefer to remain in one place.

They are less prone to get into territorial conflicts since they are more willing to play and nurture their young. The male Emperor Tamarin is very protective of his female companion during mating and will defend her from rival males.

In addition, he will protect her, feed her, and keep her near him. On the other side, the female is less possessive and often departs from the male to go exploring or foraging for food. Emperor Tamarins often reproduce in the rainy season, when it is warmer outside and there is more food available.

The male will show off his beautiful silver mane, groom the female, and feed her during this period of courtship. The couple will mate as soon as the female is receptive, and after a gestation of 173 days, the female will give birth to a single child.

Male and female Emperor Tamarins are quite different from one another in several ways, some of which include size, color, behavior, and style of life, to name just a few of these distinguishing features. Male and female Emperor Tamarins have very different lives than one another.

The lifestyles of male and female Emperor Tamarins are quite different from one another. It is important not to lose sight of the reality that reproduction from both sexes is required to guarantee the survival of the species, even if it is fascinating to learn about the unique qualities that are carried by each gender.

CHAPTER FOUR

Advantages

An Emperor Tamarin's alluring look is perhaps its greatest perk. It has a distinctive appearance due to its longer-than-its-body black and white mustache. It enhances the tamarin's individuality in addition to being aesthetically pleasing.

Being a highly sociable creature, Emperor Tamarin enjoys interacting with its owners. Additionally, it is a fairly clever species that can pick up basic orders and skills. Emperor Tamarin may develop into a devoted and obedient companion with regular and appropriate training.

The Emperor Tamarin is also a very simple animal to maintain. Fruits and vegetables, which may be acquired from pet shops or online, make up the majority of the tamarin's diet. They do not need a lot of room and may live well in relatively tiny cages.

Furthermore, because the rainforest is the tamarin's native home, keeping a warm, humid climate is not difficult. And last, Emperor Tamarin may be maintained in a variety of settings since it is a highly adaptable species.

They can survive in both indoor and outdoor habitats and are not fussy eaters. The Emperor Tamarin may be a lovely addition to any household as long as they have a safe and secure place to call home.

Disadvantages

Emperor Tamarin is a wild animal and might be challenging to manage as such. If not handled appropriately, they may be unpredictable and could bite or scratch. Additionally, they are unsuited for flats or other tight living spaces due to their tendency to be quite loud and energetic.

The price is just another drawback to having an Emperor Tamarin. They may be costly to maintain and need for a particular diet. Additionally, they may be expensive to construct or buy and need a large, safe enclosure.

The Emperor Tamarin thrives in social environments and is best kept with other tamarins. While some people may not have an issue with this, it might be difficult for others without the resources to find a good home for many tamarins.

Similar breeds

To make sure the breed is a suitable match for your lifestyle and house before choosing a pet, do your homework. Due to its tiny size, gregarious behavior, and distinctive look, the Emperor Tamarin is a preferred option for pets.

While Emperor Tamarin is a rare breed, there are a number of comparable breeds that might make excellent pets for certain people.

The Cotton-top Tamarin is a little monkey that is indigenous to the tropical woods of Panama and Colombia. The Cotton-top Tamarin is a petite, sociable animal that enjoys playing and interacting with its people, much like the Emperor Tamarin.

The distinctive white mane of fur on Cotton-top Tamarins covers their whole body from head to shoulders. They are also said to be very clever, and they can be taught to identify their owners and obey basic instructions.

Second, another little monkey that is sometimes contrasted with

Emperor Tamarin is the Black-capped Squirrel Monkey.

Their intelligent, playful, and endemic Central and South American monkeys are well known for their qualities. Black-capped Squirrel Monkeys are gregarious animals that may develop close relationships with their owners, much like Emperor Tamarin. They need plenty of enrichment and action to maintain their health and happiness since they are also highly active.

Third, another little monkey species that resembles Emperor

Tamarin is the White-faced Saki Monkey.

These intelligent, gregarious monkeys are endemic to South American tropical woods and are prized for their intelligence. They differ from other tiny monkeys because of their black body and white face.

White-faced Saki Monkeys may still be a fantastic option for pet owners searching for a distinctive and clever companion, even if they can demand a little more attention from their owners than Emperor Tamarin.

Fourth, another kind of little monkey that resembles Emperor Tamarin is the Pygmy Marmoset. These primates, who are endemic to the tropical woods of South America, are renowned for their intellect and fun natures.

Pygmy Marmosets are very little, making them a fantastic option for those seeking a pet that will remain diminutive. Pygmy Marmosets need a lot of enrichment and action to keep fit and happy, much as Emperor Tamarin does. The Golden Lion Tamarin is a tiny species of monkey that is unique to Brazil's tropical jungles.

These monkeys are gregarious, and they may develop close relationships with their owners, much like Emperor Tamarin. Additionally intelligent, they may be trained to recognize their masters and obey directions.

Golden Lion Tamarins are distinguished from other tiny monkeys by their lovely golden fur, which is very well recognized. In conclusion, there are a number of possibilities available if you are seeking a pet that resembles Emperor Tamarin. The Emperor Tamarin is comparable in size, intellect, and social behavior to the Cotton-top Tamarin, Black-capped Squirrel Monkey, White-faced Saki Monkey, Pygmy Marmoset, and Golden Lion Tamarin. F

or owners searching for a distinctive and perceptive pet, any of these kinds of monkeys may make wonderful companions.

Conclusion

One of the most fascinating and distinctive creatures in the world is the Emperor Tamarin. To think that such a little creature can play such a significant role in our life is amazing.

It is hard to avoid being charmed by this extraordinary species, whether it is because of Emperor Tamarin's outstanding physical features or its lively personality. In its natural environment, Emperor Tamarin is a significant species and a keystone species.

It contributes significantly to preserving the well-being of its immediate surroundings. Unfortunately, it is currently considered an endangered species as a result of deforestation and other human-related activities. It is up to us to do something to support the preservation of this species habitat.

We should appreciate and admire the extraordinary species known as Emperor Tamarin. It is a particularly remarkable animal because of its distinctive appearance and habits.

From its adaptation and tenacity to its playful personality and intellect, Emperor Tamarin can teach us a lot. We may learn a lot about ourselves and how we connect with the world around us by observing Emperor Tamarin. We must take the necessary action to save this species and its environment.

In addition to ensuring its life, doing so would enable us to keep enjoying its positive effects in our lives. The fascinating animal and significant component of our planet are Emperor Tamarin. Its distinctive traits and actions serve as a reminder of how diverse life is on Earth.

To safeguard the survival of Emperor Tamarin and its environment, we must all do our part. Together, we can make sure that this amazing species survives and continues to enrich our lives.

THE END

Printed in Great Britain
by Amazon

24082127R00046